W9-DDB-133

Royal
Gardens
Royal

Claire Masset

Buckingham Palace
A Royal Garden

ROYAL COLLECTION TRUST

Contents

Introduction

Buckingham Palace is one of the world's most famous buildings. A symbol of the British monarchy, it is the official London residence of Her Majesty The Queen. Every day, people gather in front of its impressive gates to soak up the prestige and pageantry of this unequalled setting. Some visitors come to see the Changing of the Guard; others may be lucky enough to catch sight of a horse-drawn royal carriage.

Throughout its history, Buckingham Palace has been the scene of many public celebrations. Royal weddings and coronations, as well as national events such as VE Day, have been marked by the Royal Family appearing on the palace balcony to greet waiting crowds.

The formality of the palace's façade, gloriously complemented by the Queen Victoria Memorial, stands in stark contrast to its secluded garden. Hidden from sight, it is first and foremost The Queen's private London garden. Like its owner, however, it fulfils many roles. Every summer it welcomes thousands of guests for the famous Garden Parties. It has played host to rock concerts and sporting events and regularly serves as a helicopter pad.

Beyond these private and public functions, the garden is a green lung at the centre of one of the world's biggest and busiest cities. As well as displaying the finest horticulture, it harbours an astonishing diversity of plants and wildlife, which in themselves are an ecological treasure.

Within its 16 hectares (39 acres), Buckingham Palace Garden offers a unique array of beautiful and remarkable plants, features and views. Its broad sweep of lawn gives way to a lake set against a magnificent treescape. Circling this open space are sinuous paths edged with dense shrubberies, flower-filled borders and romantic meadows – all home to a growing collection of plants. Set in the lake is a secluded island, itself a miniature landscape and a haven for birds and other wildlife.

The gardeners at Buckingham Palace balance a nature-friendly approach with the very best horticultural practices. This book takes you on a tour of the garden over the course of a year, showing how it grows and develops through the seasons. Along the way, you will enjoy gardening tips from Head Gardener Mark Lane, as well as insights into the garden's long and rich history.

Things have changed considerably since James I planted his Mulberry Garden near the site of the present garden in the early 17th century. It is pleasing to know that this historic feature inspired the creation of a National Collection of Mulberries in 2000. And while the garden caused a sensation in the 18th century when it housed one of the first zebras ever seen in England, there is now similar excitement as rare and unusual flora and fauna are being discovered within its walls.

Plan of Buckingham Palace Garden

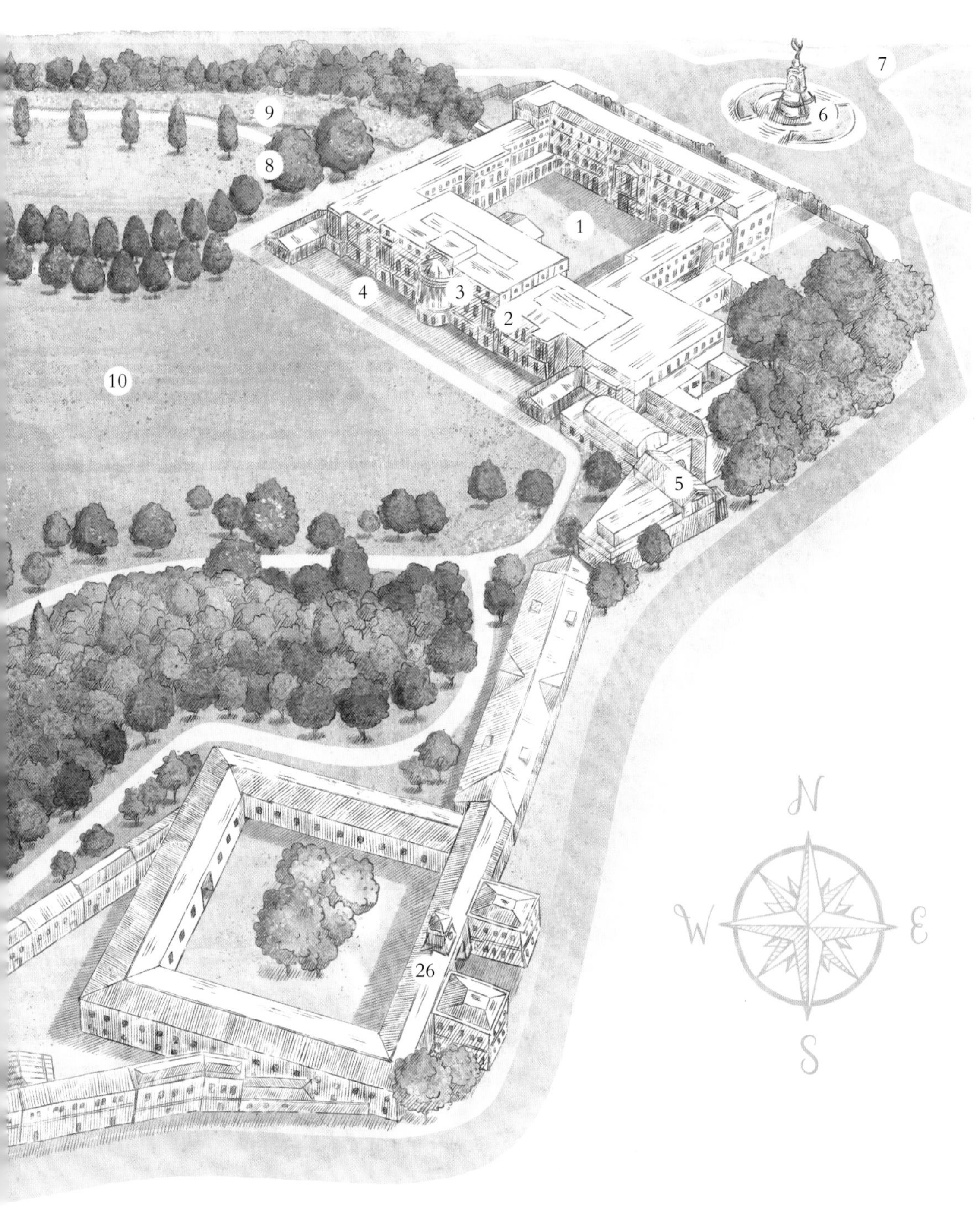

KEY

1. Buckingham Palace
2. West Front
3. Bow Room
4. West Terrace
5. The Queen's Gallery
6. Queen Victoria Memorial
7. The Mall
8. Victoria and Albert Plane Trees
9. Herbaceous Border
10. Main Lawn
11. Queen's Walk
12. Round Bed
13. Camomile Lawn
14. Bronze Cranes
15. Lime Tree Roundel
16. Rose Garden
17. Admiralty Summer House
18. Waterloo Vase
19. Magnolia Dell
20. Wellington Arch
21. Recycling Centre
22. Waterfall
23. Beehives
24. The Mound
25. Greenhouse
26. Royal Mews

Spring

Spring arrives early at Buckingham Palace, thanks to its sheltered position in the heart of London.

Signs of emerging life are everywhere: in the budding shrubs, the fresh leaves of the lakeside willows and the carpets of pale primroses. Like birds during the dawn chorus, the garden starts to sing.

As the weeks progress, the velvety buds of magnolias burst open to reveal their spectacular blooms. Blossom sprinkles the garden with snowflake petals and drifts of wildflowers warm the heart as they set the scene for a late spring crescendo of azaleas and rhododendrons. Meanwhile, the gardening work steps up in preparation for the summer parties.

PREVIOUS PAGE: Before the London plane trees come into leaf, the Angel of Peace, atop the Wellington Arch on Hyde Park Corner, is framed elegantly from the garden.

OPPOSITE: The view from the island towards Buckingham Palace.

LEFT: With its magnificent deep pink blooms on bare branches, *Magnolia* 'Vulcan' enjoys pride of place at the front of the Magnolia Dell.

BELOW, CLOCKWISE FROM TOP LEFT: *Magnolia* 'Elizabeth' in bud.

Weeping willows soften the contours of the water's edge.

Cherry blossom can range from pink to pure white.

Many different cherry species grow throughout the garden.

The garden's beautiful flower-spangled meadows are among its most delightful spring spectacles. Many of the flowers have spread naturally over time – primroses, daffodils and bluebells in particular – creating splendid carpets of colour. Every year, though, the gardeners cannot resist introducing a few new additions. As a result, the garden now harbours many different types of daffodil. One of the most distinctive is dainty *Narcissus cyclamineus*, whose bright yellow petals stand dramatically upright, as if windswept.

Although romantic and painterly, the meadows are not just designed to please the eye. Covering 2 hectares (5 acres) in total and home to more than 320 different wildflowers and grasses, they are biodiversity hotspots, supporting a rich variety of insects, from bees and beetles to grasshoppers and butterflies. And while some of the meadows are tucked away in the wilder parts of the garden – on the lake edges and in woodland glades – a few are close to the palace itself, breaking up the formal areas of lawn. These large expanses of open meadow hark back to the years when George III and Queen Charlotte turned parts of the garden to pasture – a time when cows and goats grazed the grass, which was no doubt speckled with daisies, buttercups and other wildflowers.

TOP RIGHT: Related to onions, *Allium paradoxum* var. *normale* forms dense leafy clumps and sends up bell-like flowers that are popular with bees and have a garlic aroma.

BOTTOM LEFT: Tazetta daffodils usually carry several flowers on a single stem, but this one is particularly generous, with at least eleven.

BOTTOM RIGHT: Small and dainty, *Narcissus cyclamineus* is immediately recognisable thanks to its backswept petals.

ABOVE: Pheasant's eye daffodils (*Narcissus poeticus*) stand tall in the meadows in late spring, their rich fragrance wafting on the breeze.

BOTTOM LEFT: Although they look exotic, snake's head fritillaries (*Fritillaria meleagris*) are native to the UK. The pinkish-purple flowers have a distinctive chequerboard pattern.

BOTTOM RIGHT: Bluebells spread happily beneath the trees, enjoying the dappled spring sunlight before the leafy canopy forms overhead.

Queen Charlotte's menagerie

Queen Charlotte, the wife of George III, adored animals and in her time the garden was home to many unusual creatures, not just cows and goats. In 1762 she ordered the vegetable patch to be transformed into a paddock for 'the Queen's animals'. Among them were a female zebra and an elephant. Elsewhere there were aviaries filled with unusual birds, and two monkey houses. This weird and wonderful menagerie attracted much public attention and crowds flocked to gawp at the zebra, one of the first ever seen in England. One observer recalled: 'The Queen's she-ass was pestered with visits, and had all her hours employed from morning to night in satisfying the curiosity of the public.' The famous artist George Stubbs even painted her portrait, which – like its real-life model – caused quite a sensation. Queen Charlotte also had a menagerie near her cottage at Kew Palace, where kangaroos – yet another novelty for the British public – were introduced in the 1790s.

Zebra, painting by George Stubbs, 1763

Since 2008 the garden has been home to five beehives. They are nestled in a secluded spot on the island that is – according to the beekeeper who looks after them – 'near-heaven for bees'. Filled with wildflowers, grasses and flowering shrubs, this uncultivated haven is as natural a habitat as you can get in the city. The simple wooden hives are positioned so that the bees take off over the lake, away from summer party guests. The bees are a relatively gentle species, *Apis mellifera ligustica*, also known as the Italian honey bee.

Thanks to the garden's many winter flowers, there is ample food to sustain the bees over the colder months. As the temperature warms and more flowers appear, the bees become more active. The honey is harvested in June or July, when the beekeeper judges the time is right. The hives produce about 160 jars of clear, golden honey per year, which is used in the royal kitchens.

OPPOSITE, LEFT: Looking towards the island's beehives as one of the resident coots glides past.

OPPOSITE, TOP RIGHT: Spring blossom provides essential pollen and nectar for the garden's many bees, including its colonies of honey bees.

OPPOSITE, BOTTOM RIGHT: Simple and traditional, these wooden beehives are regularly checked by the beekeeper to ensure their inhabitants are healthy and thriving.

RIGHT: Busy worker bees of the *Apis mellifera ligustica* species – *mellifera* meaning 'honey bearing' in Latin.

PREVIOUS PAGE: Snake's head fritillaries, both purple and white forms, have naturalised in the grass near the Round Bed.

LEFT: This hybrid strawberry tree (*Arbutus × andrachnoides*) is one of the original trees planted on the Mound.

BELOW: The garden's Pulhamite rocks harbour many types of plants, from small trees and shrubs to wildflowers, and are covered in mosses and lichens.

BOTTOM: Spring starflowers (*Ipheion uniflorum*) dot the Mound with upward-facing, honey-scented blooms.

In the southern part of the garden, a large area known as the Mound is awash with daffodils, bluebells and other spring bulbs. This secluded spot is thought to originate in part in the early 19th century, when William IV ordered his gardeners to bring in material from outside the garden to create the beginnings of a high mound. Its construction helped to mask the view of the recently erected Royal Mews and today it still provides precious privacy for the gardens.

In 1904 Pulhamite rocks were installed on the southern side of the Mound to create a naturalistic 'rockscape' that was planted with shrubs and trees. A Victorian invention, Pulhamite is a mix of sand, crushed brick and cement over a base of rubble, designed to look like stone. Extremely fashionable in the 19th and early 20th centuries, Pulhamite was used to create rockeries, ferneries, grottoes and even temples and fake cliffs.

Here at Buckingham Palace, these famously durable rocks have stood the test of time. Small trees and shrubs still grow among them, including maples, buddleias, forsythias and hollies. Meanwhile, the rocks' damper, flatter surfaces are the perfect terrain for lichens and mosses.

Changes in fashion

The garden you see today looks nothing like it did in the 18th century, when it reached a peak of formality. The then gardener, Henry Wise, added lots of finely executed features, including a canal (shown on the right side of the painting), avenues of lime trees, elaborate parterres with fountains and statues, and paths edged with neatly spaced tubs of bay and orange trees.

As was the case with so many other great English gardens, these formal elements were stripped away when the fashion for the more naturalistic landscape style took hold in the mid- to late 1700s. When George III and Queen Charlotte started residing at the palace in the 1760s, the garden entered a new, almost bucolic phase. The canal was filled in, parterres were dug up and much of the old ornamental garden turned into rolling landscape.

Buckingham House, painting attributed to Adriaen van Diest, c.1703–10

In late spring, a very special flower comes into bloom. The white helleborine (*Cephalanthera damasonium*) is a native orchid whose numbers are declining, yet it was discovered growing in the garden in 2013. Seven plants were found on the banks of the lake during a London-wide botanical survey. This elegant, shade-loving orchid usually prefers the chalk and limestone woodlands of southern England and had not been seen in London since 1900.

Today the colony is thriving and has spread to other parts of the garden, including the Mound. Significantly, the white helleborines at Buckingham Palace have flowers that regularly open fully, something that happens rarely in wider national and European populations.

The garden harbours many other rare and unusual native plants and no doubt more will be discovered in the future. Treasures include the diminutive silver hair grass (*Aira caryophyllea*), which is seldom seen in London, as well as the pretty hare's foot clover (*Trifolium arvense*), with its hair-covered pink flowers that look like little paws.

At this time of year the lake edges become a sanctuary for nesting waterfowl. Many of the birds construct their nests in sheltered, out-of-the-way spots, but coots are not so discreet. Theirs are large, untidy, mound-like constructions of reeds and twigs that appear to float on the lake. Soon the island and surrounding water become nurseries for young birds, some of which – particularly the goslings – venture onto the Main Lawn to graze.

OPPOSITE: A white helleborine (*Cephalanthera damasonium*). This native orchid starts blooming in either late spring or early summer.

ABOVE: A coot's nest sits on the lake as the sunrise creates elegant silhouettes and watery reflections.

LEFT: Two Egyptian geese and their goslings feeding on the lawn near the sundial. The garden is also popular with greylag and Canada geese.

OPPOSITE: Primrose-yellow *Magnolia* 'Elizabeth' was bred at Brooklyn Botanic Garden, which gave this fine tree to The Queen in 1982.

ABOVE: Robust and versatile, varieties of *Camellia* × *williamsii* often have a very long flowering season, from late winter through to early summer.

Over the years, many resident kings and queens have appreciated the garden's spring-flowering shrubs and trees. Queen Victoria notes in her diary on 13 May 1843: 'It was so fine in our pretty garden, with all the azaleas & rhododendrons out'. In April 1844 she talks of 'all the lilacs coming out & the apple trees loaded with blossom'. On other occasions, she delights in the cherry and almond blossom, the camellias and the laburnums.

Queen Mary, wife of King George V, was also fond of spring flowers and blossom, and her son, King George VI, and his wife, Queen Elizabeth, adored flowering trees. During the 1930s and 1940s the couple oversaw the planting of magnolias, cherries and camellias, many of which still thrive today.

Spring-flowering trees are peppered throughout the garden, but some areas – such as the Magnolia Dell and Queen's Walk – are filled with them, creating large and particularly rich displays. Very early in the season, sometimes even in late winter, the camellias come into bloom. There are more than 200 different varieties in the garden, ranging from single flowers to frothy, peony-like blooms, and from pure white through to mottled pinks and intense reds. Their dark, glossy foliage shows off the blooms to perfection.

Once the camellias have finished their first flush of flowers, the magnolias start their display. These elegant trees and shrubs bear large, often fragranced blooms that open before the leaves, producing a dramatic effect against bare branches. Flowers vary in form from upright and tulip-shaped to open and star-like. Most are in delicious shades of pink and white, but a few are pale yellow, such as the variety 'Butterflies', a gift to The Queen for her Golden Jubilee in 2002.

TOP ROW, LEFT TO RIGHT: The delicate, ever-so-slightly pink blooms of *Prunus × schmittii*.

Magnolia × soulangeana. The magnolias in the garden range in colour from pure white to dark maroon and every shade in between.

Newly opened blooms of *Magnolia* 'Peppermint Stick' have slender, upright petals.

Camellia japonica 'Lavinia Maggi' is one of a number of camellias in the garden with striped flowers.

BOTTOM ROW, LEFT TO RIGHT: 'Black Tulip' is prized for its dramatic flowers, among the darkest of any magnolia.

Some cherries have light, airy blossom; others form densely packed clusters that provide maximum impact.

Viburnum carlesii 'Diana' is a compact shrub producing abundant flowers with a sweet, spicy scent.

Azaleas and rhododendrons are the stars of the late spring garden. When seen *en masse* their abundant flowers produce intense tapestries of scarlets, mauves, oranges and yellows. Some display more delicate shades of pink, cream and white, offering a romantic counterpoint to the intensity of their neighbours.

Thanks to an extensive programme of border renovation, the garden now enjoys freshly planted beds of medium-sized azaleas and hybrid rhododendrons. These add a striking middle layer to spring's floral canvas. A few of these shrubs, such as the bright yellow azalea *Rhododendron luteum*, emit enticing scent.

The garden's most historically significant rhododendron is arguably 'London Calling'. A cross between fragrant white 'Loderi King George' (named after King George V) and vivid red 'Bulstrode Park', this beautiful pink rhododendron was raised in the garden and presented to The Queen on her 70th birthday in 1996.

OPPOSITE: The garden's oldest evergreen azaleas (*Rhododendron* 'Amoenum') produce clouds of magenta blooms, even in light shade. The most striking display of azaleas and rhododendrons can be found along Queen's Walk, which leads to the Rose Garden.

ABOVE: *Rhododendron* 'London Calling' was bred by the current Head Gardener.

FOLLOWING PAGE: Edged with stately plane trees, this curving path takes you past the Magnolia Dell and leads to the magnificent Rose Garden just round the corner.

The Queen's rose

Many flowers have been named after members of the Royal Family. Just a few – such as *Rhododendron* 'London Calling' – were specially raised at Buckingham Palace, but many originate from around the world. Possibly the most popular of all royal blooms, *Rosa* 'Queen Elizabeth' was bred in the United States by the famous rose grower Dr Walter Lammerts. It was created to celebrate The Queen's Coronation in 1953 and became immediately fashionable with the rose-growing public. It is tall and extremely hardy – eminent rose specialist David Austin described it as 'indestructible' – and its pure pink, upright blooms are held on strong, almost thornless stems, making them ideal for cutting. This classic rose still grows in the garden today.

Rosa 'Queen Elizabeth', watercolour by Lotte Günthart, 1961

Seasonal activities

Maintaining the lawns

- As soon as the grass starts growing in spring, it is mown weekly to keep it at the right height. Frequent mowing creates a dense, healthy lawn and reduces weeds.

- The lawn edges are clipped regularly to add precision to the overall look of the garden. It's a slow but worthwhile task, as neat edges set off all the other features beautifully.

- Stripes are created in formal areas of lawn using a mower with a built-in roller. They help to lead the eye and make the garden or lawn look larger, as well as framing the flower borders.

Mulching

● Spreading a layer of mulch, about 5 cm (2 in) deep, over the soil surface helps to suppress weeds and retain moisture. Applying a little fertiliser beforehand also gives the plants a much-needed spring boost.

● Composted bark is the mulch of choice in many of the shrubberies, as it has a low pH, which many of the plants prefer. It also has a light, open texture, so allows moisture through when watering.

● Camellias thrive on a liberal application of mulch, but the gardeners are always careful to leave a gap around the base of each plant, as direct contact with mulch can rot the stems.

Summer

As summer dawns, the garden is poised to welcome its guests. Garden Party season has arrived.

Traditionally, The Queen hosts three Garden Parties a year, usually in May. This is an opportunity to thank people from all walks of life for their outstanding work. Months of preparation go into making each party an occasion to remember. Each is attended by around 8,000 guests, invited at the recommendation of Lord-Lieutenants, local and national government, the Services, religious groups, charities and societies.

Guests enter through one of three gates and catch their first glimpse of the garden: a great swathe of lawn with sunlight reflecting from the distant lake. With its magnificent wooded backdrop, this is the perfect setting for a special afternoon. Two large marquees line the southern side of the lawn. Guests are served from the tempting displays of sandwiches and cakes. Military bands add to the festive atmosphere, performing familiar tunes, from recent hits to classical marches.

At 4pm sharp, The Queen emerges from Buckingham Palace accompanied by other members of the Royal Family. When she reaches the steps at the top of the West Terrace, silence falls as the National Anthem is played. The party then continues for another two hours, by when 27,000 cups of tea, 20,000 sandwiches and 20,000 slices of cake will have been consumed by guests who will savour the occasion for the rest of their lives.

PREVIOUS PAGE: Golden morning light filters through the dense canopy of the Indian chestnut avenue.

OPPOSITE: Buckingham Palace bathed in soft golden light at daybreak.

RIGHT: The Queen and The Duke of Edinburgh arrive to greet their Garden Party guests.

The first Garden Party

Queen Victoria is famous for her long period of mourning after Prince Albert's death in 1861. After more than six years of seclusion, on 22 June 1868, she hosted the first ever Garden Party at Buckingham Palace. 'The afternoon splendid, & not too hot,' she records in her journal. 'Quantities of people on the lawn whom I had to recognise as I went along […] it was very puzzling & bewildering.' Despite her anxiety, the day was a resounding success, as was recorded in *The Illustrated London News*. Two particularly lavish Garden Parties were held to celebrate Queen Victoria's Golden and Diamond Jubilees in 1887 and 1897.

The Garden Party at Buckingham Palace, 20 June 1887, painting by Frederick Sargent, 1887–9. Queen Victoria can be seen centre left, accompanied by the Prince of Wales (later King Edward VII).

Although the Main Lawn is the focal point of the parties, guests are free to explore the whole garden and take in its many features and different areas – from the formal beds and open lawns to wilder, woodier and more secluded spots.

A few guests may come across Buckingham Palace's famous camomile lawn, which gives off a deliciously sweet, apple-like scent as you walk over it. Camomile (*Chamaemelum nobile*) was first recorded here in 1666. Despite its historical importance, it requires no special treatment; it forms part of the Main Lawn and is therefore looked after in the same way.

As soon as Garden Party season is over, the lawn undergoes an intense maintenance regime to ensure it is in tip-top condition for the palace's summer opening, which usually starts in late July. Once the marquees are taken down, bare and damaged patches of lawn are resown with grass seed and covered with horticultural fleece. This holds in warmth and moisture, while also providing protection from birds. Given these ideal growing conditions, the seeds start to germinate within two weeks. Soon the lawn is once again green and lush, ready for visitors – and the resident geese – to enjoy.

OPPOSITE: Not far from the Main Lawn, the garden takes on a wilder aspect, with weeping willows and other mature trees and shrubs creating restful green canopies.

LEFT: Greylag geese are easily distinguishable by their grey-brown plumage, orange bill and pink legs.

Close to the palace's West Terrace, the avenue of Indian chestnuts (*Aesculus indica*) is one of the garden's great formal features. These tall, broad trees look especially glorious in June, lighting up the garden with candle-like clusters of small white flowers, tinged pink and flecked with yellow.

Elegantly curved, the avenue was planted in 1967 to provide a secluded walk from the palace to the wooded areas of the garden. The trees' flowers produce lots of much-needed food for bees and other pollinators at a time when – perhaps surprisingly – many gardens experience a 'June famine' and insects struggle to find enough food.

The garden's many wildflower meadows also benefit pollinators during this period. Along the lake edge and under trees, meadow plants have been allowed to naturalise. Different areas offer different conditions – from sunny and dry to wet and shady – encouraging a greater variety of flowers and wildlife to thrive.

The garden's population of wildflowers is constantly being increased and currently totals around 250 different species. The Lime Tree Roundel, for instance, is a recent re-creation of a long-gone feature. It comprises a horseshoe-shaped curve of small-leaved lime trees (*Tilia cordata* 'Winter Orange'), beneath which a meadow, supplemented with pollinator-friendly alliums, native geraniums and salvias, is slowly maturing.

OPPOSITE (BOTH): The Indian chestnut avenue reaches its peak in June, with a magnificent show of candle-like flower clusters.

TOP LEFT: Attractive to bees and hoverflies, the flowers of mouse garlic (*Allium angulosum*) turn from pale to dark pink as they mature.

TOP RIGHT: The recently planted Lime Tree Roundel, with its curved row of small-leaved lime trees.

BOTTOM LEFT: Drumstick alliums (*Allium sphaerocephalon*) have naturalised in one of the meadows.

As you reach the north corner of the palace, two large London plane trees come into view. Planted by Queen Victoria and Prince Albert, they have come to symbolise the royal couple. No one knows who planted which tree, but it is pleasing to speculate. As a pair – with their broad canopies reaching up and touching one another – they are a living, growing memorial to the couple's love, strength and unity.

There are plane trees throughout the garden, as single specimens and in clusters and avenues. This is entirely fitting: the London plane (*Platanus* × *hispanica*) – as its common name suggests – is widely grown in the city, making up more than half of its tree population. But this was not always the case here at Buckingham Palace. Fifty years ago, the garden was home to about 100 English elms (*Ulmus procera*), which formed the backbone of the treescape. By the 1980s every single one had succumbed to Dutch elm disease and plane trees have since taken precedence. The garden now has 98 mature plane trees.

London planes are ideal for city-centre conditions, coping with air pollution thanks to their self-shedding bark. However, their fallen seeds germinate very readily, which creates a challenge for the gardeners, who need to be on constant watch for seedlings. If left unchecked, the garden would become a forest of plane trees.

OPPOSITE: Known as Victoria and Albert, the garden's two most famous plane trees were planted by the Queen and her consort more than 150 years ago. They now stand as a living memorial to the royal couple.

Painting the garden

A keen and talented watercolourist, Queen Victoria enjoyed painting the trees at Buckingham Palace and in other royal gardens, such as Windsor and Balmoral. In her journal, she notes her delight at seeing leaves emerge in spring and witnessing their autumnal tints. On 13 May 1851, while in residence at Buckingham Palace, she writes: 'The garden looks so pretty just now, with all the trees out in their tender fresh green, & the lilacs in bloom.'

ABOVE: *A view from a window at Buckingham Palace*, watercolour by Queen Victoria, June 1853

RIGHT: *Tree in Buckingham Palace Garden*, watercolour by Queen Victoria, late April 1847

Buckingham Palace Garden, watercolour by Queen Victoria, 25 June 1857

The impact of the Herbaceous Border rests not only in its size – 156 m (512 ft) long by 5 m (16 ft) deep – but in the huge diversity of plants and their colourful and dramatic juxtapositions. There is very little in the way of repetition here; every step offers a new delight.

In early spring, the perennials in the border start to wake from their winter slumber. As the season progresses, they slowly bulk up and grow taller. By summer the plants at the back reach such heights that visitors can be fooled into thinking that the border is planted on a slope rather than on flat ground. It rises towards you, like a colourful wave.

At its peak, the border ripples with colour and texture, brimming with all manner of flowers – from stately delphiniums and romantic phloxes to warm-coloured daylilies (*Hemerocallis*) and heleniums. Sculptural exotics such as bananas and cannas, along with bold beauties like plume poppies (*Macleaya*), bring drama, while the dark green backdrop of trees and shrubs creates a perfect foil for all the colour. Every year a few new exotics and tender plants, such as the blue potato bush (*Lycianthes rantonnetii*), are added to the border, further increasing its wow factor.

Fifteen sweet-pea wigwams, evenly spaced along the length of the border, are the one concession to formality. But here too there is diversity, as each wigwam plays host to a different variety of sweet pea. Every year the Head Gardener chooses a new set of sweet peas in collaboration with The Queen's Royal Florist, who uses them in flower arrangements for the palace and in The Queen's posies (see pages 99 and 112).

PREVIOUS PAGE: An impressive 156 m (512 ft) in length and packed with a rich assortment of flowers, the Herbaceous Border is one of the glories of the garden.

BELOW: Thanks to the garden's mild city-centre climate, these banana plants survive the winter with a simple covering of straw.

OPPOSITE, TOP: This section of the border features heleniums, helianthus, phlox, buddleia, Russian sage and deep purple *Salvia nemorosa* 'Plumosa'.

OPPOSITE, BOTTOM ROW: All the sweet peas are grown from seed at the Royal Gardens, Windsor, before being sent to Buckingham Palace for planting out in spring.

TOP ROW, LEFT TO RIGHT: A highly architectural plant, *Acanthus mollis* produces spectacular spires of white and purple flowers.

Persicaria amplexicaulis forms mounds of long, pointed leaves topped with wands of pinky-red flowers that are particularly attractive to bees.

With its soft bottlebrush plumes, feathertop grass (*Pennisetum villosum*) adds graceful touches to the Herbaceous Border.

The tall spikes of *Veronica longifolia* 'Blauriesin' bloom all summer long.

BOTTOM ROW, LEFT TO RIGHT: With rich crimson flowers, *Lilium* 'Claude Shride' is one of the most elegant Turk's cap, or martagon, lilies.

Like all phlox, fragrant 'Mount Fuji' is adored by bees, butterflies and other pollinators.

The blue potato bush (*Lycianthes rantonnetii*) can reach 2 m (6½ ft) tall, adding drama, colour and height to the Herbaceous Border.

As you make your way along Queen's Walk, you come across the sweetest of surprises: the Rose Garden, truly the jewel in the crown of this magnificent garden.

Exquisite and immaculate, it is made up of formal beds, very much in the Victorian style. This garden, however, dates from the 1960s and is the creation of garden writer and rose expert Harry Wheatcroft, who did much to popularise rose growing among British gardeners in the mid-20th century.

Sixty rose bushes grow in each of the 25 beds and each bed contains a different variety of rose, chosen for its fragrance, colour and disease resistance. The garden is designed in such a way that no two adjacent beds feature a similar colour, allowing each rose to show off its individual charms.

From the neat lawn edging to the rich, weed-free soil, attention is paid to every detail, so that the rose bushes are always beautifully presented, healthy and flower abundantly for as long as possible. They are pruned in November – slightly earlier than is the norm – and generously mulched in spring to ensure they are in flower for the first Garden Party guests in May. And they continue to bloom throughout the summer, guaranteeing a fragrant spectacle for all who visit.

OPPOSITE: The freshly cut, neatly striped lawn is the perfect foil for the neat rose beds.

TOP: The large-flowered, deep crimson *Rosa* 'Rob Roy'.

BOTTOM: 'Tickled Pink' is prized not only for its delightful name but also for its masses of healthy, lightly scented, medium pink blooms.

FOLLOWING PAGE: The Rose Garden in full bloom.

Most of the roses in the Rose Garden are hybrid teas. These are strong, disease-resistant and repeat-flowering plants that produce elegant buds and large blooms, proudly held on upright stems. Colours in the beds range from yellows ('Eurostar') through to apricots ('Special Occasion'), pinks ('Lovely Lady') and scarlets ('Remembrance'). Whether admired individually or as a group, the flowers have an old-world charm all of their own, harking back to a time when gardeners believed roses were so precious they deserved a spot of their own in a garden.

To maintain the Rose Garden's flawless condition, every year one of the rose beds is completely replanted. The bed is dug to a depth of about 75 cm (30 in) and replaced with fresh soil to avoid the risk of rose replant disease. 'Mum in a Million', with sumptuous pink flowers, is one of the most recent arrivals, and there are plans to reintroduce a few standard roses – a trained form on a tall single stem – which were part of Harry Wheatcroft's original design.

Beyond the formal Rose Garden are two large rose beds planted in a looser style. One of these is home to the golden-yellow rose 'Golden Wedding', given to The Queen and The Duke of Edinburgh by members of the public for their Golden Wedding Anniversary in 1997.

As one might expect, roses with royal connections can be found throughout the garden. Rich red 'Royal William', created to mark Prince William's birth in 1982, is here in the Rose Garden. 'Elizabeth of Glamis', named after Queen Elizabeth The Queen Mother and a beautiful blend of pink and salmon, can be seen on the North Front Lawn, across from The Queen's Apartments. The apricot-peach 'Silver Jubilee' has been in the garden, appropriately, since 1977.

OPPOSITE, LEFT: Climbing rose 'New Dawn'.

OPPOSITE, RIGHT: The striking *Rosa* 'Mamma Mia'.

TOP: The bright and memorable 'Eurostar' rose.

RIGHT: Super-floriferous *Rosa* 'Remembrance' produces dense clusters of vivid red blooms.

Within the Rose Garden are two historically significant features. While one is relatively discreet, the other impresses by its size. Dominating the centre of the lawn is the Waterloo Vase. Carved from three pieces of Carrara marble, it weighs an estimated 19 tonnes and is an impressive 5.5 m (18 ft) tall. Meanwhile, the Admiralty summer house is a picture of daintiness and refinement. It nestles on the edge of the rose beds, clad in a double-flowered purple wisteria, 'Violacea Plena'. Adorned with four statues of sea gods, this early 18th-century building originally stood on the old Admiralty site at the other end of The Mall. It was given to King Edward VII and Queen Alexandra, who brought it to the garden in 1906.

BELOW: Facing each other in the Rose Garden, the Waterloo Vase and the Admiralty summer house are two of the garden's most exquisite features.

A vase with a story

Had all gone to plan for the French Emperor, Napoleon, this magnificent vase would have ended up in France. Confident of his success in battle, and intending to commemorate his glorious victory, he commissioned it prior to setting out on his Russian campaign in 1812.

Napoleon's defeat at Waterloo in 1815 meant the collapse of his empire, after which the part-completed vase was presented to the Prince Regent. In 1820, by then King, George IV asked sculptor Richard Westmacott to finish the vase.

Westmacott's carved reliefs show horses and men sweeping across the vase in battle scenes, with Wellington and Napoleon portrayed as victor and vanquished and the figure of Europe taking refuge at the throne of England. The vase proved too heavy to be housed in Windsor Castle and in 1835 William IV presented it to the nation. First placed in the National Gallery, the vase was brought to the garden at Buckingham Palace in 1906.

This photograph is from a series taken by Cecil Beaton of Queen Elizabeth, later The Queen Mother, in 1939. The Queen Mother is wearing a dress from the White Wardrobe, a range of 30 outfits designed for her by the leading fashion designer of the day, Norman Hartnell.

Queen Elizabeth photographed in the garden at Buckingham Palace by Cecil Beaton, 1939.

TOP ROW, LEFT TO RIGHT: With their graceful bowing branches, weeping willows are one of the most distinctive and elegant trees along the water's edge.

One of two rusticated bridges leading onto the island. Both are made from Pulhamite rock (see page 27).

From midsummer, globe thistles (*Echinops ritro*) produce striking silvery-blue flowerheads with a bounty of easily accessible nectar.

The garden's picturesque waterfall also serves a practical purpose: it helps to circulate water in the lake, thereby oxygenating and revitalising it.

BOTTOM ROW, LEFT TO RIGHT: Coots build their prominent, raised nests near the water's edge, using nearby vegetation such as reeds and grasses.

A meadow of ox-eye daisies (*Leucanthemum vulgare*). One of the most recognisable of all wildflowers, its single, open blooms are highly attractive to pollinating insects.

Seasonal activities

Harvesting the mulberries

- The mulberry harvest takes place in June and July. It lasts about three to four weeks, as the fruits don't all ripen at the same time. The gardeners go out weekly to gather the darkest, ripest berries.

- Each picking session can take a couple of hours – harvesting must be done slowly and carefully to make sure the delicate fruits don't get damaged.

- The harvested mulberries are used in the royal kitchens in special seasonal desserts, such as apple and mulberry crumble and mulberry soufflé.

Deadheading roses

- Regularly removing all the fading blooms from the rose bushes is a key priority and ensures the Rose Garden looks in tip-top condition throughout the summer.

- Rather than simply snipping off a spent flower, the gardeners cut the stem further down, just above the first or second leaf below the old bloom. A new flower bud on a brightly coloured shoot then sprouts from that point in a couple of weeks.

- Deadheading not only keeps the rose bushes looking immaculate, but also stimulates more flower buds to grow, extending the display.

Autumn

In autumn the garden undergoes a dramatic transformation. Its usually green backdrop becomes a mesmerising centrepiece, as leaves slowly start to glow.

For a glorious few weeks, the scene is aflame with radiant colours, from bright yellows and ambers to fiery oranges, reds and coppers. Every day brings new shades to the fore. A yellow-stemmed ash (*Fraxinus excelsior* 'Jaspidea') lights up one of the paths with its clear yellow foliage. Stag's horn sumachs (*Rhus typhina*) and tupelos (*Nyssa sinensis* and *Nyssa sylvatica*) radiate golds and burnt oranges, while berberis show off some of the brightest scarlets in the garden.

Smoke trees hold their colour for a satisfyingly long time. Among them is the luminescent *Cotinus coggygria*, whose leaves turn vibrant dark purple. Maples include the brilliant red *Acer rubrum* 'October Glory'. Some trees, not least the liquidambars, display an intense range of colours all at once.

A handful of notable specimens keep their green leafy canopy for a few extra weeks. This is the case with the Victoria and Albert plane trees (see pages 48 and 49): a special type of London plane, they were selected because they hold their leaves for longer. Today they still perform their strategic role of screening the palace.

PREVIOUS PAGE: Japanese maple *Acer palmatum* 'Osakazuki'.

OPPOSITE, LEFT: This yellow-stemmed ash, *Fraxinus excelsior* 'Jaspidea', is one of the oldest trees in the garden.

OPPOSITE, TOP RIGHT: *Acer rubrum*, also known as the red maple.

OPPOSITE, BOTTOM RIGHT: Pecan *Carya illinoinensis*, a tree native to the southern United States.

TOP RIGHT: The London planes turn a rich gold.

BOTTOM RIGHT: The early autumn leaves of *Acer rubrum* are a pleasing mix of green, yellow, orange and red.

With so many different trees in its 16 hectares (39 acres), the garden is a vast arboretum. It is home to many native trees – such as English oak, white willow, hawthorn, alder, ash, beech and silver birch – but it also harbours many exotic and endangered species. Among the latter is a beautiful conifer from Chile, the plum-fruited yew (*Prumnopitys andina*), which has tasty fruit and yew-like foliage. Thanks to careful planting, monitoring and maintenance, the gardeners have been able to grow trees that are rarely seen in the UK, such as the Chinese chestnut (*Castanea mollissima*) and the round-leaved beech (*Fagus sylvatica* 'Rotundifolia').

Over the last 30 years the garden has been planted with variety in mind and the range of trees has increased significantly. There are now 85 different types of oak, with unusual and interesting examples from around the world, including the Oriental white oak (*Quercus aliena*) from Japan and Korea, the deer oak (*Quercus sadleriana*) from Oregon and the Aleppo oak (*Quercus infectoria*) from southern Europe and the Middle East. Another prized oak is *Quercus* × *libanerris* 'Rotterdam', a vigorous variety bred in the Netherlands, which was planted by The Queen in 1989.

Since the early 20th century, members of the Royal Family have celebrated wedding anniversaries, coronations, births, birthdays and special visits by adding trees to the garden. King Edward VII started the trend when he planted a copper beech on the Main Lawn on 15 March 1902. Throughout his reign he became known for his ceremonial tree planting, a practice he pursued wherever he went, to the delight of the crowds.

There are 20 commemorative trees in the garden, each with a story to tell. Perhaps the sweetest relates to four English oaks. The first two of these were planted on Good Friday 1954; one by Prince Charles, from an acorn sown on the day of his birth (14 November 1948), and the other by Princess Anne, from an acorn sown on the day she was born (15 August 1950). Two other oaks were similarly raised and added to the garden to celebrate the births of The Queen's two younger children, Andrew and Edward.

OPPOSITE, TOP: Growing near the lake edge, the garden's graceful weeping willows are among the first trees to get their leaves in spring and some of the last to lose them.

OPPOSITE, BOTTOM: The garden is home to a few horticultural oddities, such as the Osage orange (*Maclura pomifera*). Despite its name and its strange, somewhat orange-like fruit, the tree is a member of the mulberry family.

RIGHT: This London plane was planted by King George V and Queen Mary to celebrate their 20th wedding anniversary. Its plaque is shown above.

FOLLOWING PAGE: A mellow autumnal dawn lights up the Main Lawn.

PAGES 76–77: In many parts of the garden, trees and shrubs seem to merge into one another, creating rich painterly effects in autumn.

The Queen's mulberries

One of the garden's most notable tree collections has an interesting historical precedent. In 1608 James I tried to develop a national silk industry in a bid to topple France's monopoly. A little north of the present palace, he planted his 1.6-hectare (4-acre) Mulberry Garden, containing about 10,000 mulberry trees. He even appointed his very own 'Mulberry Men' to tend it. Silkworms feed on mulberry leaves, but the trees James I and his men planted were black mulberries (*Morus nigra*) rather than white mulberries (*Morus alba*), which silkworms prefer, and within a few years the project failed.

Since 2000 the garden has been home to the National Collection of Mulberries, the first National Collection to be held by The Queen in London. It comprises 40 different types and – somewhat ironically – many are varieties of white mulberry, including 'Laciniata' with deeply indented leaves and weeping 'Pendula'. The collection is rich and varied, featuring tall and thin (fastigiate) to dwarfing and non-fruiting forms, and two varieties of red mulberry (*Morus rubra*), whose leaves turn vivid yellow in autumn. Mulberry fruits are delicious eaten raw or in pies and are harvested in June and July (see page 66).

Morus nigra (black mulberry), detail of a watercolour from the florilegium by Alexander Marshal, *c.*1650–82. Despite producing deliciously juicy fruit, this type of mulberry does not generally appeal to silkworms.

OPPOSITE, TOP ROW, LEFT TO RIGHT:
Sweet-smelling, late autumn-flowering *Mahonia* × *media*.

Casuarina cunninghamiana, commonly known as the river she-oak.

Camellia sasanqua 'Navajo' flowers throughout autumn and sometimes into winter.

OPPOSITE, BOTTOM ROW, LEFT TO RIGHT:
Yellow-berried holly, *Ilex aquifolium* 'Bacciflava'.

Male catkins of the Atlas cedar (*Cedrus atlantica*).

The strawberry tree (*Arbutus unedo*) produces strawberry-like fruits that ripen to red in late autumn.

The Comus Room, illustration from *The Decorations of the Garden Pavilion in the Grounds of Buckingham Palace*, 1845, a printed volume produced under the supervision of Prince Albert's art adviser Ludwig Grüner.

A private retreat

Queen Victoria and Prince Albert saw the Mound – with its lake views and sense of remoteness – as the perfect spot for a garden pavilion. Designed as a 'place of refuge', this cottage-like building was completed in 1845 to designs by the architect Edward Blore, who was at the time rebuilding parts of the palace.

We have Prince Albert to thank, however, for the building's exquisite interiors. He was Chairman of the Royal Commission on the Fine Arts, which had recently been appointed to oversee the decoration of the new Palace of Westminster. The pavilion was a trial run for this grand project and the result was remarkable. Along with gilded doors, a splendid fireplace and all manner of friezes and vignettes, the central Comus Room featured frescos by distinguished artists, including Edwin Landseer, Victoria's favourite painter. Another room was inspired by the novels of Sir Walter Scott, much enjoyed by Albert, while a third was decorated in the Pompeiian style. Every surface was an opportunity for ornament.

Perched atop the Mound, the pavilion was reached by open carriage along gently sloping paths. Here the royal couple could escape from courtly life or enjoy small gatherings. Sadly, the pavilion fell into disrepair during the First World War and was taken down in 1928.

As autumn progresses, the garden's many evergreens really come into their own. Large holly and bay trees, glossy camellias, bushy rhododendrons and spectacular conifers give structure to the garden at a time when most trees and shrubs drop their leaves. As well as providing a pleasing green framework, they help to maintain the garden's intimate atmosphere.

One spot where this is particularly apparent is on the Mound, which was partly designed to protect the garden and palace from unwanted views and noise. In the Victorian era, great masses of native holly (*Ilex aquifolium*) were planted here to give extra shelter. Today this type of work carries on: trees and evergreen shrubs are being added both on the Mound and elsewhere to ensure the garden's continued peacefulness.

Trees may be the main autumn attraction, but there are also plenty of flowers to enjoy. Although it has passed its summer peak, the Herbaceous Border remains colourful. Among its late blooms are elegant Japanese anemones, cheerful helianthus and exotic ginger lilies.

Flowering until the first frosts and loved by pollinators, salvias are the stars of autumn, both here in the border and elsewhere in the garden. A particularly impressive variety is *Salvia* 'Amistad', which produces showy spikes of appropriately royal purple blooms. Salvias come in a wide range of often striking flower colours – from yellows and pinks to blues, purples and crimsons – and, although the plants are short-lived, they are easy to propagate from cuttings. The gardeners at Buckingham Palace have recently introduced even more of these sun-lovers, with a focus on eye-catching types such as *Salvia sclarea* var. *turkestanica*, with tall pinky-white flower spikes, and *Salvia leucantha*, with felty purple flowers and light, airy growth.

In the wilder parts of the garden, tiny autumn cyclamen (*Cyclamen hederifolium*) form carpets of pretty pink flowers that last well into winter. They naturalise readily and are happy growing in shady corners and among the roots of large trees. Their attractive heart-shaped leaves continue beyond flowering time.

Perhaps the most delicate blooms are the autumn-flowering camellias. These graceful beauties come in shades of pale pink, deep pink or white, set off against the handsome backdrop of evergreen foliage. They form a lovely link with the garden's many spring-flowering camellias (see page 31).

TOP ROW, LEFT TO RIGHT: Autumn blooms in the garden include *Elsholtzia stauntonii*, a range of pretty salvias and *Anemone hupehensis*.

BOTTOM ROW, LEFT TO RIGHT: *Cyclamen hederifolium* creates cheerful patches of autumn colour in some of the shadier parts of the garden.

Camellia 'Snow Flurry', one of the many camellias that form the backbone of the autumn show.

FOLLOWING PAGE: The garden and palace are a perfect counterpoint to one another, each enhancing the other's beauty. Here, a stag's horn sumach (*Rhus typhina*) and other trees frame the view of the palace from the island.

Autumn brings yet another cheering sight: a rich bounty of colourful berries, from traditional bright reds and oranges to pinks and blacks. The most unusual-looking are those of the spindle tree (*Euonymus*), whose reddish-purple fruits split open to reveal bright orange seeds, forming a curious yet pleasing contrast. Nutrient-rich berries are a vital food source for birds during the winter months and are particularly loved by song and mistle thrushes, blackbirds and visiting redwings and fieldfares.

In an average year, the garden plays host to over 50 different types of bird, of which about 30 are resident. Among the nesting species are wrens, dunnocks, robins, blackbirds and thrushes; visiting birds include green and great spotted woodpeckers, treecreepers and sedge warblers.

Buckingham Palace: gardens, lake and Garden Pavilion, watercolour by Caleb Robert Stanley, *c.*1845. This view of the lake hints at the variety of birdlife during Queen Victoria's reign – notice the charming little bird houses set in the water to the left and right.

Queen Victoria's birds

During their years in residence, Queen Victoria and Prince Albert acquired all kinds of rare and aquatic birds for the garden. Victoria delighted in their quirky features and characters, which she records on several occasions in her journal. On 15 March 1841 she writes: 'Our birds, consisting of pheasants, ducks, Goodwits [*sic*], pigeons, a dear little red-legged partridge, &c are a great amusement to us; but our tamest amongst them is a Gannet or Solan Goose […] He is the tamest thing I ever saw; when he is quite far off, at a distance, the moment he sees us, he waddles along as fast as he can, after us, & if he is on the water, he swims after us.'

OPPOSITE, CLOCKWISE FROM TOP LEFT: Washington thorn, spindle bush, holly and Father David's rose are some of the brightest jewels adorning the autumn garden.

TOP: A flock of Canada geese gathers under the Indian chestnuts.

INSET: One of the garden's many bird boxes. This one regularly provides a home for broods of great tits.

RIGHT: Rich red smoke bushes (*Cotinus coggygria*) add drama to the water's edge.

The lake is a magnet for water birds and forms part of their 'London circuit'. Geese, ducks, swans, herons and cormorants come and go between Regent's Park, Kensington Gardens and St James's Park, unhindered by any physical boundaries. Overall, there is a higher concentration of birds in Buckingham Palace Garden than in many other parts of London.

Trees and shrubs provide food, cover and protection not just for birds but for lots of other creatures too, from bats and beetles to moths and squirrels. Even old and dead trees play a unique role in supporting a wide range of species. Throughout the garden are several 'standing totems' – dead trees that the gardeners have made safe by removing any hazardous branches. Their fissured bark and many cracks and holes offer precious habitats for birds, bats, moths and about 300 different types of beetle.

Similarly, old tree stumps are left *in situ* to rot down, attracting beetles, wasps, hoverflies and fungi. Dead wood is full of nutrients that sustain a host of creatures and, as the wood decomposes, it enriches the soil. Woodpile stacks on the Mound and island are another vital habitat, particularly favoured by solitary bees and ladybirds.

With more than a thousand trees of different sizes in the garden, leaf collecting is a mammoth task in autumn. Leaves are gathered up and taken to the Recycling Centre for composting (see page 108). The regime is slightly different on the island, where fallen leaves are simply added to the continually evolving piles of wood and sticks, forming yet another wildlife haven.

OPPOSITE, LEFT: The trunk of an old Chinese chestnut (*Castanea mollissima*).

OPPOSITE, RIGHT, TOP TO BOTTOM: Piles of logs and branches make ideal shelters for small creatures. Many types of beetle take up residence in dead and decaying wood, as these bore holes suggest, and all manner of fascinating fungi thrive in the garden.

INSET: Grey squirrels spend much time foraging on the ground in autumn.

RIGHT: A carpet of oak leaves.

Even during Queen Victoria's time the island was looked after differently from the rest of the garden. Wilder, shadier and generally more overgrown, it acted as a refuge for nesting birds. This remains true today. In fact, the island is now a rich and finely balanced ecosystem: an oasis within an oasis.

To help maintain this natural environment, gardeners go onto the island as little as possible. Birds and other wildlife remain largely undisturbed, and the mix of meadows, lakeside plantings, dense shrubberies and trees offers many different habitats. Protective evergreens help to create a microclimate on the island. Being surrounded by water, it also has a more humid environment than the rest of the garden, encouraging different flora and fauna. Recent surveys on the island have revealed unexpected and exciting finds, including two rare beetles (*Longitarsus ferrugineus* and *Clitostethus arcuatus*), as well as a fungus (*Cristinia coprophila*) not recorded here since 1938.

PREVIOUS PAGE: On the island the vegetation is so dense that one can easily forget the garden's central London location.

TOP: Common reeds (*Phragmites australis*) line the water's edge, providing ideal nesting and hiding places for waterfowl.

OPPOSITE: One of the best Japanese maples for autumn colour, *Acer palmatum* 'Osakazuki' eventually turns a brilliant scarlet.

Seasonal activities

The autumn tidy-up

- One of the biggest tasks in the gardeners' calendar is collecting all the leaves that fall from the many deciduous trees and shrubs in autumn. The team clears the lawns first, before meticulously working through the shrubberies and flowerbeds.

- Herbaceous plants naturally die back in autumn, so to keep the borders looking neat, all the faded top growth is cut down in October and taken to the garden's Recycling Centre for composting.

- After a busy summer of use, the lawns are in need of repair. Fertiliser is applied, while any worn patches and larger areas of damage, perhaps caused by a marquee, are reseeded. The turf soon recovers in the mild, damp autumn weather.

Planting trees and shrubs

- Autumn is an ideal time to plant deciduous trees and shrubs, as they get the chance to settle into their new home before spring growth starts.

- The roots of pot-grown plants often become congested. Loosening or breaking up the outside of the rootball before planting encourages new roots to grow outwards, helping the plant to get established.

- As the roots spread into the surrounding soil, the plant is able to take up the water and nutrients it needs, resulting in healthy fresh growth.

Winter

Denuded of much of its greenery, the winter garden takes on a different but no less alluring character.

The landscape – in all its contours and undulations – reveals itself and lets in glimpses of the outside world. As you walk around, views that are disguised for most of the year become apparent, offering new and unexpected perspectives.

The changing weather has a great impact on the garden's atmosphere and appearance. On overcast days, steely grey skies have a magic of their own – a charm redoubled by their reflections in the lake. As soon as the sun comes out, it lights up the pale birches and fiery-stemmed dogwoods, while the tracery of plane tree branches comes into warm focus against the blue sky. Snow is rare in London these days, but even a dusting of frost is enough to transform the garden into a shimmering wonderland.

PREVIOUS PAGE: Looking towards Buckingham Palace from the island.

OPPOSITE: A crisp winter morning bathes the palace in bright sunlight, while glassy reflections on the lake double the impact of the treescape.

RIGHT: Snow is a rare occurrence in central London, but when it does fall the garden is enveloped in a wonderfully peaceful atmosphere.

While your eye is first drawn to the wide vistas, charming details slowly come to the fore. Characterful tree bark, small scented flowers and adorable garden birds – suddenly conspicuous on the newly bare branches – take on a starring role.

Stripped of their leaves, the garden's tree skeletons show off a variety of textures, colours and shapes. Among these are paperbark maples (*Acer griseum*), whose cinnamon-coloured bark peels in strips, revealing yet deeper colours beneath. In contrast, the snake-bark maples (*Acer capillipes*) flaunt smooth olive-green stems with pale vertical stripes. Dotted around the garden are white-stemmed birches – bright beams among the shrubberies – and, by the lake, false acacias (*Robinia pseudoacacia*) show off their contorted forms.

Perhaps the most delightful feature of the winter garden is fragrance. Happening upon the scent of a winter-flowering honeysuckle (*Lonicera fragrantissima*), wintersweet (*Chimonanthus praecox*) or daphne is a spirit-lifting experience. Since 2001 the gardening team has been planting more of these small-flowered gems, adding yet further to the enchantment.

Even during the coldest months, the garden has floral interest, as proven by the constant supply of blooms for The Queen's posies. The tradition of creating a posy for The Queen started in 1992. Every Monday, a small bouquet featuring about half a dozen blooms is placed in a vase on Her Majesty's writing table. This little floral tribute is a lovely way of bringing the outdoors in, offering a snapshot of the garden. In winter, a posy might consist of an early flowering camellia, a few snowdrops and hellebores, together with sprigs of *Viburnum tinus*, winter jasmine (*Jasminum nudiflorum*) and heather.

OPPOSITE, TOP: The garden is home to a diverse collection of trees and shrubs, both evergreen and deciduous.

OPPOSITE, BOTTOM ROW, LEFT TO RIGHT: A conspicuous winter robin.

Colourful willow stems beside the lake.

Japanese cornelian cherry (*Cornus officinalis*).

TOP ROW, LEFT TO RIGHT: Paperbark maple (*Acer griseum*), river birch (*Betula nigra*) and strawberry tree (*Arbutus unedo*).

BOTTOM: This dawn redwood (*Metasequoia glyptostroboides*) grew as a stunted specimen for many years, creating an unusual bulbous base.

Top row, left to right: The waxy, bell-like flowers of *Clematis urophylla* 'Winter Beauty' open from December through to February.

Japanese cornelian cherry (*Cornus officinalis*) produces clusters of delicate yellow flowers in late winter.

Daphnes are some of the sweetest-smelling shrubs in the garden.

Wintersweet (*Chimonanthus praecox*) lives up to its name – its small, almost translucent flowers are deliciously fragrant.

Bottom row, left to right: The yellow-flowered witch hazel, *Hamamelis* × *intermedia* 'Pallida', another of the garden's many winter-scented shrubs.

Viburnum tinus forms a large evergreen shrub, its dark glossy foliage contrasting with white or pale pink flowers.

Oregon grape (*Mahonia aquifolium*) thrives in dense shade, flowering from late winter in this sheltered setting.

During the winter, the palace building itself – along with other sculptural features in the garden – holds the gaze more than at any other time of year.

Even on the greyest of days, the West Front's honey-coloured stone radiates warmth. The gracefully curved Bow Room stands out beautifully, while classical-style reliefs, vases and urns add decorative flourishes to the façade, itself a picture of architectural harmony.

Beside the lake, two bronze cranes cast their own elegant silhouettes. Created in Japan in the second half of the 19th century, they were brought to the garden from Osborne House, on the Isle of Wight, and displayed in a formal area near the palace. They were later moved – at the behest of King George VI – to the water's edge, where they look very much at home. A common feature in Japanese art and gardens, cranes symbolise good fortune, peace and harmony. Their tranquil setting here seems entirely fitting.

PREVIOUS PAGE: Weeping willows are interspersed with stately conifers on the lake edge.

LEFT: Looking across to the famous Bow Room from the other side of the lake.

TOP: One of two friezes that run across the palace's West Front. The carved stone in this bas-relief shows Alfred the Great expelling the Danes from England.

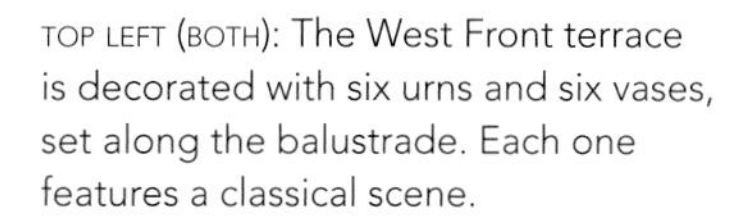

TOP LEFT (BOTH): The West Front terrace is decorated with six urns and six vases, set along the balustrade. Each one features a classical scene.

BOTTOM LEFT: Two large 19th-century bronze cranes grace the waterfront.

BELOW: The garden's lamp posts are topped with a small gilded coronet.

By late winter the garden starts to show signs of fresh growth. In the wild grass areas, masses of snowdrops and bright yellow winter aconites make their appearance. Sweet violets and cyclamen add colour to woodland areas and, by the water, the weeping willows come into radiant leaf, filling the lakeside with a green-gold glow.

These ancient willows date back to the 1880s and are now sadly in decline. Only five of the original trees remain, but as one slowly fades, a new sapling is planted nearby. Eventually, these young trees will grace the garden with their own gently weeping boughs.

BELOW: Great swathes of lawn are speckled with snowdrops in late winter.

Skating at Buckingham Palace, February 1895, from photographic albums compiled by Princess Victoria of Wales (daughter of King Edward VII)

Winter sports

Queen Victoria and Prince Albert loved the great outdoors and enjoyed the cold weather. Frost and snow were always welcome, providing opportunities for skating parties and sledging fun. 'After luncheon went again to see the children sledging in the garden. It is a delightful thing & these real winter amusements & real winter weather are most enjoyable,' Victoria records in her journal on 16 February 1853. Later that month, she writes: 'The ice has been so strong and thick that there has been delightful skating.'

Albert was an accomplished skater, but that did not stop him suffering the odd accident. 'I managed, in skating, three days ago, to break through the ice,' he writes, in a letter to a friend in 1841. 'I was making my way to Victoria, who was standing on the bank with one of her ladies, and when within some few yards of the bank I fell plump into the water, and had to swim for two or three minutes in order to get out. Victoria was the only person who had the presence of mind to lend me assistance.'

ABOVE, LEFT TO RIGHT: Snowdrops, *Crocus tommasinianus*, *Cyclamen coum* and sweet violets (*Viola odorata*).

Work in the garden does not slow down in winter. This is a busy time for planning and maintenance, but also for planting, leaf gathering, seed sowing and even lawn mowing. Warmer weather patterns have seen the lawns continue to grow through the winter, necessitating several cuts during the season.

By far the greatest winter task is leaf collecting, which takes four months to complete. Clearing leaves from the lawns with leaf-blowers is a lot more straightforward than cleaning out dense shrubberies, where leaves get stuck among the plants and can require hand-picking.

Since the turn of the century, a small percentage of leaves are deliberately left on the ground. This policy change was introduced following research revealing that blackbird numbers were declining across the Royal Parks, because parks and gardens were too tidy. Many small birds like to flick leaves over, revealing tasty bugs underneath. Today their numbers in the garden are once again on the increase.

Another relatively recent introduction was the Recycling Centre, installed in 1991 at The Queen's request. Ninety-nine per cent of all green waste from the four royal gardens in London – Buckingham Palace, Kensington Palace, St James's Palace and Marlborough House – is recycled here. The remaining one per cent comprises pernicious weeds and diseased material, which always need to be burnt. The compost produced in the Recycling Centre is used to enrich the soil, either dug in or laid on the surface as a mulch.

ABOVE: In the Recycling Centre, grass clippings, leaves and other garden waste is composted, then used as a soil improver in the garden.

Thanks to the gardeners' hard work, by winter's end the garden is ready for the cycle of the seasons to begin again. As fresh growth starts to emerge, the garden comes to life once more.

Seeds of success

A favourite activity during cold and wet winter weather is seed sowing, which takes place in the palace's greenhouse (see page 113). Come spring, the resulting seedlings are planted into the vegetable garden at nearby Clarence House, the official London home of The Prince of Wales and The Duchess of Cornwall. No vegetables are currently grown at Buckingham Palace, but this was not the case during the First World War, when some of the flowerbeds were turned to vegetable production.

Vegetables being harvested in the garden during the First World War, photographs from one of Queen Mary's photographic albums, 1918

TOP ROW, LEFT TO RIGHT: The garden is visited by more than 50 different types of bird, including blackbirds, green woodpeckers and jays.

ABOVE: Paperbark maple (*Acer griseum*) underplanted with a ground-covering, starry-flowered periwinkle, *Vinca difformis*.

FOLLOWING PAGE: Winter gives way to spring as daffodils and magnolias start to flower, adding intrigue and gaiety to the scene.

Seasonal activities

Creating winter posies

- When The Queen is in residence, the gardeners create a weekly posy of cut flowers from the garden (see page 99). During winter, as an alternative to flowers, the posies can feature a mix of evergreen leaves and colourful berries.

- Using clean, sharp secateurs, the gardeners cut long stems that will stand well in a vase. It's best to wear gloves when harvesting prickly favourites such as holly.

- Choices from around the garden range from berry-rich cotoneaster and mistletoe to hawthorns and crab apples, as well as rosehips and dried seedheads.

In the greenhouse

- During winter, the gardeners clean out and disinfect the greenhouse and thoroughly wash all the pots, seed trays and labels for reuse in spring. They also regularly check the greenhouse plants to ensure they are pest free.

- From late winter, seeds start to be sown into small pots. The resulting seedlings are thinned out to leave just one or two in each pot. Once large enough, they can be transplanted into open ground with minimal root disturbance.

- After sowing the seeds, the gardeners look after them carefully to ensure they get the right amounts of water, warmth and light to germinate and grow well.

A miscellany of flowers, seeds and fruits through the year

TOP ROW, LEFT TO RIGHT: *Stachyurus praecox* 'Issai' produces catkin-like clusters of pale yellow flowers in February and March.

The ripening berries of *Fatsia japonica* turn a deep black as they mature. Featuring large and glossy palmate leaves, this evergreen shrub has a particularly exotic appearance. Given the right conditions, it can easily be grown outdoors.

Witch hazels display their distinctive spidery flowers all through winter, in shades of yellow, orange and red, like this one, *Hamamelis vernalis* 'Carnea'.

The garden's many London planes produce these spherical fruits made up of a dense cluster of individual seeds.

BOTTOM ROW, LEFT TO RIGHT: The striking seedhead of Chinese liquorice (*Glycyrrhiza uralensis*), a flowering plant whose roots are widely used in Chinese medicine.

Catkins adorn the common alder (*Alnus glutinosa*) from late winter to early spring. The male catkins are pendulous, up to 10 cm (4 in) in length, whereas the female ones are shorter and develop into woody cones, as can be seen in the top right of this image.

Callicarpa rubella, also known as beautyberry because of its jewel-like, purple fruit.

Timeline

1608

James I plants his Mulberry Garden near the present site of Buckingham Palace.

1640

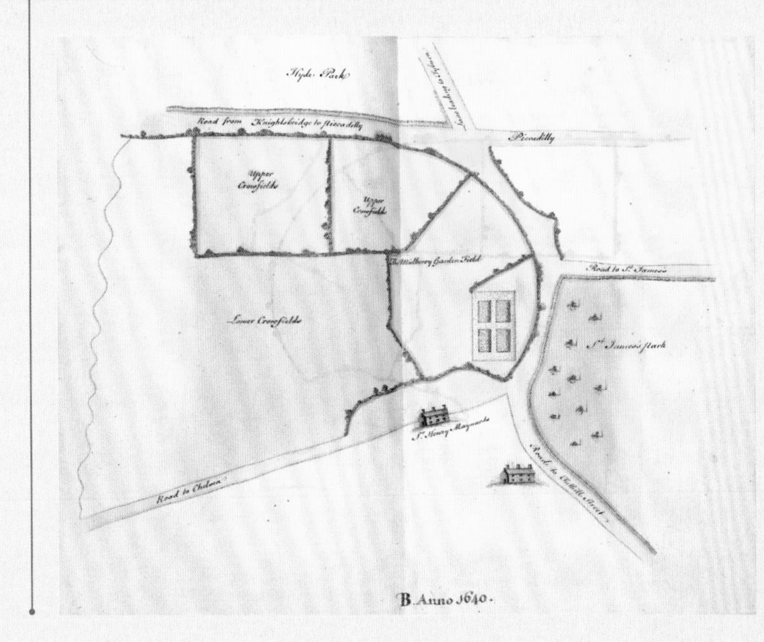

Lord Goring acquires the Mulberry Garden, followed by other land around his mansion, Goring House. His 11-hectare (27-acre) garden becomes known as the Goring Great Garden.

1663

The Earl of Arlington acquires Goring House. He rebuilds the house and spends lavishly on the gardens.

1698

John Sheffield, Duke of Buckingham, rebuilds the house – now Buckingham House – and employs Henry Wise, the greatest gardener of the day, to lay out a formal garden.

1762

George III and Queen Charlotte acquire Buckingham House, which becomes known as The Queen's House.

1820

George IV transforms Buckingham Palace. William Townsend Aiton landscapes the grounds in the 'natural' style.

1837

By the time Queen Victoria moves in, the garden is in a sorry state, but under Prince Albert's management it quickly improves.

1897

Queen Victoria hosts a grand Garden Party to celebrate her Diamond Jubilee.

1940

The Blitz starts on 7 September; Buckingham Palace is hit nine times.

1950s

A swing, sandpit and slide for Prince Charles and Princess Anne are installed near the North Terrace.

1960s

Rose specialist Harry Wheatcroft plants the Rose Garden.

2002

The Party at the Palace takes place in the garden to celebrate The Queen's Golden Jubilee. Acts include Paul McCartney, Elton John and Annie Lennox.

2014

The Queen's Garden, a two-part documentary presented by Alan Titchmarsh, is filmed over the course of the year.

2018

The Queen's Green Planet features The Queen and Sir David Attenborough walking in the garden as they discuss The Queen's Commonwealth Canopy project.

2020

Lemon verbena, hawthorn berries, bay leaves and mulberry leaves from the garden are used as ingredients in a new gin released by Buckingham Palace.

Acknowledgements

We are grateful to Her Majesty The Queen for permission to produce this book.

In putting together *Buckingham Palace: A Royal Garden* we have been assisted by many of our colleagues from across the Royal Household. Our thanks firstly go to Mark Lane, Claire Midgley-Adam and the wonderful gardening team at Buckingham Palace, whose hard work and dedication maintains this beautiful, nature-friendly oasis in the heart of London.

Thanks are also due to Joanna Stickler, Tony Wheeler, Alexandra Little and Divya Patel, who all helped make the photo shoots possible; Jackie Fergusson and Charlotte Martin; and the staff of the Royal Archives, in particular Julie Crocker.

From Royal Collection Trust we are grateful to Curators Kate Heard and Helen Trompeteler; Katie Buckhalter, Press Officer; Karen Lawson and Daniel Partridge of Photographic Services; and Polly Atkinson, Hannah Bowen and Georgina Seage from the Publishing team.

Lastly, we would like to thank experts Tim Freed, David Darrell-Lambert and John Chapple, who answered questions about moths, birds and bees respectively.

Claire Masset would like to thank Mark Lane for his expertise, always imparted with humour and kindness; John Campbell, for all the early mornings and late evenings; Jean Postle, for her astonishing attention to detail; Michael Keates, for his impeccable visual flair; Julia Zott, for creating the delightful plan of the garden; Karen Constanti, for designing the timeline; and Tom Love, as enthusiastic and dedicated an editor as one could wish for.

John Campbell would like to dedicate his contribution to this book to the memory of Gillian Campbell.

Picture credits

Royal Collection Inventory Numbers (RCINs):
Page 27: 404350; page 45 (top): 407255; page 49 (top left): 981313.bc, (top right): 981678, (bottom): 980031.ce; page 63: 2315149; page 75: 924407; page 78: 708005; page 84 (bottom): 919889; page 107 (top): 2924736; page 108 (bottom left and right): 2303827.a–b; page 116 (bottom left): 918908, (top right): 400146; page 117 (top left): 405286, (top right): 8531507.

Royal Collection Trust is grateful for permission to reproduce the following:

Page 23 (top left): © Yale Center for British Art, Paul Mellon Collection

Page 28: © TravelMuse / Alamy Stock Photo

Pages 35 (top left), 97, 104 (left): Royal Collection Trust / © Her Majesty Queen Elizabeth II 2021. Photographer: Christopher Simon Sykes

Page 35 (bottom right): © Lotte Günthart

Page 43: Royal Collection Trust / © Her Majesty Queen Elizabeth II 2021. Photographer: Ian Jones

Page 117 (bottom middle): PA Images / PA Archive

Published 2021 by Royal Collection Trust
York House, St James's Palace
London SW1A 1BQ

ISBN 978 1 909741 69 0

102465

British Library Cataloguing-in-Publication Data:
A catalogue record of this book is available from the British Library.

Edited by Jean Postle
Designed by Mick Keates
Plan drawn by Julia Zott
Timeline designed by Karen Constanti
Photographer: John Campbell
Project Editor: Tom Love
Production Manager: Sarah Tucker
Colour reproduction: Altaimage
Typeset in Bembo and Avenir 35 Light
Printed on 150gsm Claro silk
Printed and bound in Wales by Gomer Press

Front cover: The Herbaceous Border at sunrise
(see also pages 50–1).
Back cover: 'Tickled Pink' roses (see page 57).
Page 1: Planter boxes in the greenhouse.
Pages 2–3: The Victoria and Albert plane trees.
Page 4: Circular garden seats at the base of the Victoria and Albert plane trees.
Pages 6–7: Narcissi in a wooded dell.
Pages 8–9: The changing leaves of autumn reflected in the lake.
Pages 10–11: The lush green island in summer.
Pages 14–15: Spring flowers herald the end of winter.
Page 118: The West Terrace glimpsed from across the lake.

Claire Masset is a garden writer and author of *Cottage Gardens*, *Roses and Rose Gardens* and *Secret Gardens*.

Mark Lane is Head Gardener at Buckingham Palace.

John Campbell is an award-winning photographer of gardens and natural landscapes.